Carlos Jacomino
Luis Lozano
Yazid Solís

Cybersecurity

Carlos Jacomino
Luis Lozano
Yazid Solís

Cybersecurity

challenges and solutions in digital systems to protect your information

ScienciaScripts

Imprint

Cover image: www.ingimage.com

This book is a translation from the original published under ISBN 978-613-9-41068-2.

Publisher:
Sciencia Scripts
is a trademark of
Dodo Books Indian Ocean Ltd. and OmniScriptum S.R.L publishing group

120 High Road, East Finchley, London, N2 9ED, United Kingdom
Str. Armeneasca 28/1, office 1, Chisinau MD-2012, Republic of Moldova, Europe
Managing Directors: Ieva Konstantinova, Victoria Ursu
info@omniscriptum.com

Printed at: see last page
ISBN: 978-620-8-39700-5

Introduction

In an increasingly interconnected world, where personal and corporate data is in constant digital flux, cybersecurity has become a fundamental pillar for the protection of our information. From increasingly sophisticated cyber attacks to the growing threat of vulnerabilities in critical systems, organizations and users face constant challenges in safeguarding their digital assets. This book, *Cybersecurity: Digital Systems Challenges and Solutions to Protect Your Information,* aims to provide a clear and accessible view of the risks in the digital environment and the solutions available to address them. Throughout its pages, we will explore strategies, tools and best practices to improve security in computer systems and protect the most valuable information.

In this book, you will embark on an essential journey through the world of cybersecurity, a critical discipline in the digital age. As you progress, you will discover the fundamental concepts and major threats that jeopardize the security of personal and corporate information. With a focus on defense strategies and incident response, this book offers applicable knowledge and solutions to meet the challenges of the cyber environment.

Each chapter is designed to provide you with practical tools to strengthen digital security in any context, whether personal, professional or in business environments. From the basics of cybersecurity to future trends, you will be guided in understanding threats and implementing effective preventative measures. Thus, this book will not only prepare you to understand the current landscape, but also help you to anticipate and act proactively against the risks of an interconnected digital world.

With the intention of strengthening your knowledge and raising your awareness of the importance of cybersecurity, we invite you to immerse yourself in these pages and become an advocate for safe and secure information. This learning journey will allow you to not only understand, but also contribute to the protection of systems and the preservation of trust in the digital environment.

The authors, Yazid Zahid Solis Balladares and Luis Arturo Lozano Zelaya, share their knowledge and experience in the field of cybersecurity, guiding the reader in understanding the most relevant issues in this field and presenting applicable solutions to secure the digital future of individuals and organizations.

Yazid Zahid Solis Balladares

A student of Systems Engineering at the American University of Managua (UAM), Yazid Zahid Solis Balladares has demonstrated a strong interest in the areas of Cybersecurity, software development and networking. His academic focus is on the application of innovative technological solutions to solve current challenges in the digital realm. Throughout his training, he has participated in various projects related to data protection and computer security, acquiring skills in programming, vulnerability analysis and design of secure systems.

Luis Arturo Lozano Zelaya

Also a student of Systems Engineering at the American University of Managua (UAM), he has shown a constant commitment to learning and improving in the areas of cybersecurity, computer networks and application development. His interest in emerging technologies and his hands-on approach to implementing digital solutions has allowed him to work on several information protection projects. Through his academic background, Luis has developed skills in cyber threat detection and mitigation, and is always looking to improve the security of the systems he uses.

Table of Contents

Chapter 1: Fundamentals of Cybersecurity

Cybersecurity is a critical area in today's world, where global interconnectedness and the digitization of all aspects of our lives have created an environment prone to a variety of cyber threats. This chapter explores the fundamentals that make up cybersecurity, what it means to protect information in a digital environment, and how systems are being designed to defend against cyber threats. It will discuss the most common threats and vulnerabilities, the strategies used to protect systems, and the relevance of cybersecurity in our daily lives.

What is Cybersecurity?

Cybersecurity, also known as IT security, refers to the set of practices, technologies and processes designed to protect computer systems, networks, devices and data against unauthorized access, damage or attack. As digitization and connectivity increase, organizations, governments and users must ensure that their information and digital assets are protected. From small businesses to large corporations, cybersecurity has become an essential component for operational continuity and privacy protection.

In a world increasingly dependent on technology, cybersecurity goes beyond simple data protection. It involves risk management, intrusion prevention and active defense against attacks that can compromise the integrity of IT systems. Cybersecurity encompasses areas such as network protection, information security, user privacy and resilience to cyber incidents.

Cybersecurity is crucial because the consequences of a security breach can be devastating for both businesses and individuals. The costs of a cyberattack include not only the direct loss of data or money, but also intangible costs, such as loss of trust and reputation, disruption of services and possible sanction by regulatory bodies.

In an interconnected digital world, where our lives, work and transactions are conducted online, ensuring data security has become a priority not only for businesses and governments, but also for individuals interacting in these environments.

The Evolution of Cybersecurity

The concept of cybersecurity has evolved over time, especially as computer threats have become more sophisticated. In its early days, early digital security concerns focused on the physical protection of devices and networks. With the advancement of the Internet and connectivity, attacks began to focus more on data interception and sabotage of critical infrastructure.

The rise of social networks, cloud computing and the globalization of data have given way to new threats that must be managed with advanced security technologies, such as artificial intelligence (AI), real-time data analysis and advanced cryptography. Today, cyber attacks are a constant risk for all parties involved in the exchange of digital information.

Main Elements of Cybersecurity

There are several key components that form the core of Cybersecurity:

1. **Confidentiality:** The protection of data so that only authorized persons or systems can access it. Confidentiality is essential to protect personal, business and government information.
2. **Integrity:** Ensuring that data is not altered in an unauthorized manner. Integrity is ensured through techniques such as hashing, which verifies that the data has not been modified during transmission or storage.
3. **Availability:** Ensuring that data and systems are accessible when needed, which implies maintaining a reliable infrastructure and guaranteeing that services are not interrupted by cyber attacks.
4. **Authentication and Authorization:** Systems must be able to verify the identity of users and grant them access only to resources for which they have permissions. This may include passwords, biometric authentication and access control systems.
5. **Non-Repudiation:** Ensures that a person cannot deny having performed an action in a system. Audit trails and digital signatures are tools that guarantee non-repudiation.

The world has changed radically with the advent of the digital age. What were once manual processes and siloed systems have now transformed into interconnected operations that rely on global connectivity and the ability to process large volumes of data.

Companies store their data in the cloud, banking services are performed through mobile applications, and our personal interactions are carried out on online social and messaging platforms. This digital environment has generated enormous advantages, but it has also opened the door to new threats.

Digitization brings with it the creation and massive distribution of data. Every click, transaction and movement within a system generates information that, if not properly protected, can be used for malicious purposes. This is where cybersecurity plays an essential role: protecting the information of users and organizations against unauthorized access, theft or damage.

Basic Concepts: What is Cybersecurity and why it is Crucial

In the modern digital environment, cybersecurity is much more than a technical discipline. It is a constantly evolving field that involves protecting systems, networks and data against unauthorized access, theft, damage or alteration. With the exponential growth of digitization of processes, the creation of online information and global interconnectedness, cyber threats have increased significantly, making cybersecurity crucial not only for organizations, but also for individual users.

Definition of Cybersecurity

Cybersecurity can be defined as the set of practices, technologies and processes designed to protect IT infrastructures, networks, devices and information against unauthorized access or malicious attacks. This discipline covers a broad spectrum, from the protection of personal data on social networks to the defense of critical infrastructures, such as government systems or financial services.

In simple terms, cybersecurity aims to guarantee three fundamental principles: **confidentiality**, **integrity** and **availability** of information, known as the cybersecurity triangle or CIA triad.

1. **Confidentiality:** Ensuring that data is only accessible to those who have permission to access it.
2. **Integrity:** Ensure that information is not altered in an unauthorized manner during storage or transmission.
3. **Availability:** Ensure that information is available to authorized users when they need it.

Cybersecurity in the Digital Age

The world has changed radically with the advent of the digital age. What were once manual processes and siloed systems have now transformed into interconnected operations that rely on global connectivity and the ability to process large volumes of data. Companies store their data in the cloud, banking services are performed through mobile applications, and our personal interactions are carried out on online social and messaging platforms. This digital environment has generated enormous advantages, but it has also opened the door to new threats.

Digitization brings with it the creation and massive distribution of data. Every click, transaction and movement within a system generates information that, if not properly protected, can be used for malicious purposes. This is where cybersecurity plays an essential role: protecting user and organization information against unauthorized access, theft or damage.

The Growth of Cyber Threats

Over the past decade, cyber threats have evolved from simple computer viruses to complex, high-level attacks that can cripple services, steal millions of dollars and even jeopardize national security. Cyber attacks can be both direct, such as data theft or system hijacking, and indirect, such as espionage or disruption of critical infrastructure.

Some of the most common threats include:

1. **Malware:** Programs designed to damage or access systems without the user's permission. This includes viruses, Trojans, worms and ransomware, all of which can have a devastating impact.
2. **Phishing:** Fraudulent techniques that seek to trick users into providing sensitive information, such as passwords or banking details, by pretending to be a trusted source.
3. **DDoS (Distributed Denial of Service) attacks:** Attacks that seek to make systems or services inaccessible by overloading resources with malicious traffic.
4. **Ransomware:** Malware that encrypts user files and demands

payment to restore access, which can cripple entire organizations for days.

5. **Vulnerabilities in Software:** Bugs in operating systems, applications or network infrastructure can be exploited by attackers to gain unauthorized access to systems.

As digital interactions increase, trust becomes an essential component. Individuals and organizations must trust that their data will be managed securely, and that the services they use online will protect their personal and financial information. This creates a significant liability for businesses and government entities that manage digital platforms, as a failure in cybersecurity can undermine user trust and cause irreversible reputational damage.

For example, major security breaches that have affected companies such as Facebook, Yahoo and Equifax have shown that the loss of sensitive data can have consequences not only on a financial level, but also in terms of loss of consumer confidence. Organizations must be transparent about how they handle data security and provide their users with adequate control over their personal information.

Most Common Cyber Threats

Cyber threats are diverse and constantly evolving. Some of the most common include:

1. **Malware:** Malicious software designed to damage, disrupt or steal information from systems. Among the most well-known

types of malware are viruses, worms, Trojans and ransomware. Malware attacks can have serious consequences, from data loss to the hijacking of entire systems.

2. **Phishing:** An attack in which cybercriminals attempt to trick people into revealing sensitive information, such as passwords or banking details, by masquerading as a trusted source. Phishing can be carried out via emails, text messages or fake websites.
3. **Ransomware:** A specific type of malware that encrypts files on a system and demands a payment (ransom) in exchange for the key to unlock them. Ransomware has grown enormously in recent years and represents one of the most profitable threats for attackers.
4. **Denial of Service (DDoS) attacks:** In a DDoS attack, an attacker floods a server or network with malicious traffic, causing service interruption. These attacks can paralyze websites and online services, affecting businesses and users.
5. **Vulnerability Exploits:** Uncorrected software or hardware vulnerabilities can be exploited by attackers to gain unauthorized access to systems. Exploits are one of the most common ways to compromise the security of a system.

Cybersecurity Risk Management

Risk management is a fundamental aspect of cybersecurity. It involves identifying, assessing and mitigating the risks associated with cyber threats. An effective risk management approach must consider several factors, including the potential impact of an attack, system vulnerabilities and the costs associated with security solutions.

Risk assessment should be carried out on a regular basis, as threats evolve rapidly and vulnerabilities are constantly being discovered. To mitigate risks, organizations should implement security controls, conduct regular audits and have contingency plans in case of security incidents.

Cybersecurity has come a long way since its early days. As computer technologies have advanced and the digital environment has grown, so have the threats and the solutions to combat them. To better understand cybersecurity in its current state, it is important to review its evolution, from the earliest protection systems to the complex modern approaches that require constant innovation to counter the most sophisticated attacks.

In the 1950s and 1960s, computer systems were primitive compared to today. Computers were large, expensive and mainly used by governments, universities and businesses to perform calculations and store basic information. The notion of "security" was very limited, and systems were so isolated that there was no need to worry about external threats. However, as computing expanded and became more accessible to businesses and more people, early forms of security began to take shape.

During the 1960s and 1970s, computers were mainly used in closed environments, such as universities and large institutions, and security was mainly focused on preventing unauthorized access. The first protection systems were oriented to basic authentication, such as passwords, and were used internally. In those days, the biggest threat was unauthorized access by internal users, as external networks were not a concern.

However, as computer networks began to expand in the 1980s, especially with the emergence of the ARPANET network, the precursor of what would become today's Internet, the need arose to develop better mechanisms to protect information. At that time, antivirus and firewalls began to emerge as the first lines of defense against external attacks.

The 1990s was a pivotal decade for the development of both the Internet and cyber threats. With the commercialization of the web and the rapid expansion of online users, new opportunities were created for cyber criminals. The first computer viruses, such as the famous "ILOVEYOU" in 2000, demonstrated how vulnerable information had become due to mass connectivity.

At that time, the concept of "cybersecurity" began to take on greater importance as companies began to store sensitive data online, and data protection began to be seen as a priority. Antivirus became common tools to protect personal computers against malware and viruses, while firewalls helped protect corporate networks from unwanted intruders.

The rise of the Internet brought with it new threats, such as spam, phishing and the creation of Trojans designed to steal personal information. Hackers also began experimenting with the use of vulnerabilities in operating systems and applications, leading to the development of the first security patches and automatic updates.

The Life Cycle of a Cyber Attack

To better understand how cyber attacks develop, it is essential to understand the life cycle of an attack. This cycle usually includes several phases:

1. **Reconnaissance:** The attacker gathers information about the victim, such as system vulnerabilities and weak points in the infrastructure.
2. **Exploitation:** The attacker exploits the discovered vulnerabilities to gain access to the system or network.
3. **Installation:** Once inside, the attacker can install malware or backdoors to maintain unauthorized access.
4. **Control and Exfiltration:** The attacker takes control of compromised systems and can steal data or disrupt services.
5. **Evidence Removal:** In the last stage, the attacker attempts to erase his tracks to avoid detection.

Cybersecurity Defense Tools and Techniques

There are various tools and technologies that help prevent, detect and mitigate cyber risks. Some of the most widely used include:

1. **Firewalls:** Devices or software that filter network traffic to block unauthorized access.
2. **Antivirus and Antimalware:** Software designed to detect and remove malware and viruses from systems.

3. **Intrusion Detection and Prevention Systems (IDS/IPS):** Tools that monitor networks for suspicious activity and alert administrators in real time.
4. **Encryption:** Use of mathematical algorithms to protect data during transmission and storage.
5. **Multifactor Authentication (MFA):** An authentication method that requires two or more proofs of identity, such as a password and a code sent to a mobile device.

The Future of Cybersecurity

Cybersecurity continues to evolve, and emerging technologies will play a crucial role in protecting digital systems. The use of artificial intelligence (AI) to detect anomalous patterns in network traffic, predictive analytics to anticipate attacks, and increased automation in cyber defense are just some of the emerging trends. Cybersecurity will also have to adapt to new threats, such as the impact of quantum computing, which could challenge current cryptographic algorithms.

Cyber attacks do not occur randomly or suddenly. Most cybercriminals follow a well-structured life cycle that allows them to maximize the probability of success of their attack. This cycle, which involves several phases, ranges from planning and information gathering to the execution of the attack and the subsequent exploitation of the vulnerability. Understanding this cycle is critical for organizations to effectively anticipate, detect and mitigate cyberattacks.

Public information gathering: In this phase, attackers use publicly accessible information sources, such as social networks, websites, forums and public databases, to obtain details about the targeted organization or individual. This includes information about technological infrastructures, security systems, key individuals within the organization, as well as personal details of employees, such as their job titles, locations, emails and social media connections. This information is critical to designing a specific, targeted attack.

Network and port scanning: Attackers can use network scanning tools to identify their target IP addresses and analyze open ports on servers and devices. This information allows them to identify vulnerabilities in systems, such as misconfigured services, outdated software or even publicly exposed devices. Scanning tools, such as Nmap or Nessus, are common in this phase.

Vulnerability identification: Once sufficient information has been gathered, attackers perform an analysis to identify potential vulnerabilities in the organization's systems. These vulnerabilities may include software flaws, misconfigurations, weak passwords or even flaws in security protocols. The information obtained in the reconnaissance phase is essential to determine the best way to compromise the system.

Chapter 2: Main Cyber Threats

Malware: Types and Spreading Strategies

Introduction to Malware

Malware (or malicious software) is any type of malicious program or code that affects computer systems and networks. Its purpose varies, from illicitly obtaining data to remote control of devices. The sophistication of malware propagation and infection techniques has made this threat one of the most persistent and versatile.

Main Types of Malware

Each type of malware behaves differently and affects systems in different ways:

- **Virus**: Infects executable files and spreads when these files are activated. Viruses require user action to replicate and often damage or slow down the system.
- **Worms**: They use networks to spread automatically, saturating system resources and causing crashes. Example: the **ILOVEYOU** worm, which affected millions of devices in 2000.
- **Trojans**: They hide in seemingly secure files or applications. Once the user executes them, the attacker can gain remote access to the system. Examples include **Zeus** and **Emotet**.
- **Spyware**: Designed to collect information without the user's consent. It can record keystrokes and monitor online activity.
- **Ransomware**: Encrypts files and demands payment to regain access. Examples such as **WannaCry** in 2017 had a massive impact on organizations around the world.

Malware Spreading Methods

- **Phishing Emails**: Simulate legitimate communications to trick users into downloading malicious attachments.
- **Drive-by Downloads**: Automatically triggered when visiting compromised websites.
- **Exploitation of Vulnerabilities in Software**: Exploiting security breaches in non-updated systems.
- **P2P networks and Torrents**: Malware disguised as attractive files such as music or movies.
- **External Storage Devices**: Infected USB flash drives are a common means of spread.

Impact of Malware on Systems

Malware can cause:

- **Reduced Productivity**: Slow performance and loss of time.
- **Economic Consequences**: System repair costs and loss of customer confidence.
- **Security Risks**: Access to confidential information and data integrity compromises.

Malware Attack Case Studies

WannaCry case (2017)

In May 2017, the world witnessed one of the most devastating ransomware attacks in history: the WannaCry attack. This ransomware affected more than 200,000 computers in at least 150 countries in a matter of days, crippling businesses, hospitals, universities and government agencies.

- Spread: WannaCry exploited a vulnerability in the Windows operating system known as EternalBlue, which was developed and leaked by the U.S. National Security Agency (NSA). Microsoft had released a security patch for this vulnerability months earlier, but many systems had not yet installed it, allowing WannaCry to spread rapidly.
- Mode of Operation: Once WannaCry infected a system, it encrypted all the user's files and displayed a message demanding payment of a ransom in Bitcoin to decrypt the files. The attackers demanded between $300 and $600, with a threat to double the ransom if it was not paid within a certain period of time. In addition, WannaCry used a worm technique to self-replicate and automatically spread to other vulnerable systems on the network.
- Impact: This attack had catastrophic consequences in several sectors, with the UK healthcare system (NHS) being one of the most affected. Many hospitals and clinics had to suspend critical services, cancel appointments and transfer patients due to the inoperability of their systems. In economic terms, it is estimated that the attack caused billions of dollars in losses globally.
- Lessons Learned: WannaCry underscored the importance of keeping systems up to date and applying security patches on a regular basis. It also highlighted the vulnerability of critical systems that rely on older versions of software. As a result of this attack, many organizations have strengthened their update policies and adopted data backup practices to mitigate the

impact of future attacks.

NotPetya case (2017)

A month after the WannaCry attack, in June 2017, a similar attack known as NotPetya emerged. Although it appeared to be another ransomware, NotPetya had different characteristics and targets. This attack primarily affected Ukraine, but quickly spread to other parts of the world, affecting large corporations.

- Spread: NotPetya was also spread using the EternalBlue vulnerability, just like WannaCry, but included other methods of infection, such as using stolen user credentials to access other systems on the same network. Initially, NotPetya is believed to have been introduced through a compromised update to the Ukrainian accounting software M.E.Doc, which allowed attackers to distribute the malware to a large number of systems quickly.

- Mode of Operation: Although NotPetya encrypted files on the system, its code was designed in such a way that it did not actually allow the data to be recovered, even if the victim paid the ransom. This suggests that the main goal of the attack was not financial gain, but to cause massive disruption and damage. NotPetya acted as a cyber weapon, designed to destabilize systems and networks.

- Impact: NotPetya affected global companies such as Maersk, Merck, and FedEx, causing significant disruptions to their operations. Maersk, one of the world's largest logistics companies, reported losses of up to $300 million, as the attack disabled its logistics and shipping systems for several days. In total, the NotPetya attack is estimated to have generated losses of up to $10 billion worldwide.

- Lessons Learned: The NotPetya attack was a warning about the risks of relying on third-party software without a proper security audit. It also highlighted the importance of network segmentation and the use of strong authentication, as NotPetya exploited network credentials to spread internally. In addition, he stressed the need for a disaster recovery plan to minimize the impact of catastrophic outages.

These malware cases not only exemplify the economic and operational impact that a cyber attack can have on organizations of all sizes, but also demonstrate the importance of cybersecurity in an interconnected digital world. WannaCry and NotPetya served as catalysts for many companies to review and strengthen their security policies, update their systems and adopt additional preventative measures, such as regular backups and network security hardening.

Phishing and Ransomware: The Most Common Dangers

Phishing: Social Engineering and Deception Techniques

Phishing is a method that uses social engineering to trick users into revealing sensitive information. This attack pretends to be a trusted communication (mail, SMS message, or website) to steal data such as passwords, banking information and more.

Main Types of Phishing

- **Phishing by Email**: Attackers send emails pretending to be legitimate companies. Example: emails pretending to be from banks requesting account verification.

- **Spear Phishing**: Attacks targeting specific individuals using personal information. It is more effective due to personalization.
- **Vishing and Smishing**: Phishing through phone calls and SMS messages, respectively, where the attackers pretend to be trusted institutions or entities.

Impact of Phishing on Organizations and Users

Phishing causes millions of dollars in losses every year. In 2021, phishing attacks increased significantly, compromising the security of businesses and individuals. Costs include fraud, time and resources spent on account recovery.

Ransomware: Data Hijacking and Extortion

Ransomware encrypts files on a system and demands payment (usually in cryptocurrencies) to restore access. This attack can cause total loss of critical data and disruption of operations.

Types of Ransomware

Crypto Ransomware

This is one of the most common and devastating types of ransomware. **Crypto ransomware** encrypts the victim's files, making them inaccessible. The attacker then demands a ransom to provide the decryption key to recover the files.

- **Mode of Operation**: Once it infects the system, this ransomware selects specific files, usually those with valuable information such as documents, photos, videos and work files. It uses advanced encryption algorithms (such as AES or RSA) to render the files unrecognizable and therefore unusable.

- **Notable Examples**: **WannaCry** and **CryptoLocker**. WannaCry, in particular, affected thousands of systems worldwide in 2017, including hospitals and critical infrastructure companies.
- **Impact**: Crypto ransomware can cause large financial losses, as the encrypted files are often critical to the organization's operations or the victim's personal life. In addition, recovery is difficult in the absence of backups.

2. Locker Ransomware

This type of ransomware, instead of encrypting files, locks the user's access to the entire system. The **locker ransomware** restricts access to the affected computer or device, displaying an on-screen message requesting a ransom in exchange for restoring access.

- **Mode of Operation**: Once it infects the system, this ransomware displays a lock screen that prevents the user from interacting with their computer. Often, the message warns the victim that they need to pay a ransom to regain access, and may include a timer to pressure the user.
- **Notable Examples**: One of the initial examples was the **WinLocker** ransomware, which appeared in 2011. This type of ransomware is also often seen on mobile devices.
- **Impact**: Although less damaging than crypto ransomware,

locker ransomware can prevent access to systems, affecting users and businesses that rely on the devices for their daily tasks. It is particularly problematic on devices critical to ongoing operations, such as point-of-sale.

3. Double Extortion Ransomware

This ransomware combines file encryption with a second threat: exposure of stolen data if the ransom is not paid. In recent years, this double extortion technique has become popular with attackers as it increases the pressure on victims to pay up.

- **Mode of Operation**: In the first stage, the attacker infiltrates the system, steals sensitive information and then encrypts the files. If the victim refuses to pay the ransom, the attacker threatens to make the sensitive data public or sell it on illegal markets.
- **Notable examples**: **Maze** and **REvil**. These ransomware groups have used double extortion on large organizations, leading to leaks of sensitive data.
- **Impact**: Double extortion affects not only the availability of data, but also the victim's privacy and reputation. Organizations fear the publication of sensitive data, which can cause irreversible damage to their customers' trust and public image.

4. Ransomware as a Service (RaaS)

Ransomware as a Service (RaaS) is a model in which

ransomware developers offer their malware to other attackers in exchange for a commission on the ransom. This business model has made ransomware accessible to cybercriminals without advanced technical knowledge, thus expanding the number of attacks.

- **Mode of Operation**: RaaS developers provide the infrastructure and software to conduct ransomware attacks. Affiliates" can access this technology and launch it against specific targets, sharing the proceeds with the developers.
- **Notable Examples**: **DarkSide** and **Ryuk** have used RaaS models. DarkSide, in particular, was responsible for the attack against Colonial Pipeline in 2021.
- **Impact**: This model democratizes ransomware and facilitates its distribution, increasing the volume of attacks and affecting both large and small and medium-sized companies. The RaaS infrastructure also enables organized and targeted attacks on specific sectors.

5. Non-Profit Ransomware

In some cases, attackers launch ransomware without the intention of obtaining a financial ransom. These attacks may aim to cause damage, send a message or destabilize an organization or a country. Although it behaves similarly to other types of ransomware, attackers do not provide a decryption key or means to recover data.

- **Mode of Operation**: This type of ransomware infects and encrypts files without providing a recovery path or decryption key, making the data unrecoverable.

- **Notable Examples**: **NotPetya**, launched in 2017, pretended to be ransomware, but was actually designed as a cyber-weapon to destabilize businesses and services in Ukraine.
- **Impact**: These attacks are devastating, as there is no way to recover the data, even if the victim were willing to pay. This can destabilize entire organizations and cause significant losses with no chance of recovery.

6. Scareware

Scareware uses intimidation techniques to make victims believe that their devices are infected or compromised, and demands that they pay to fix the problem. Although in many cases it does not encrypt files, scareware scares the user into paying a ransom.

- **Mode of Operation**: Through pop-up ads or fake antivirus messages, scareware pretends to have detected an infection or threat on the user's system. To resolve it, it requests a payment for a supposed "removal" of the threat.
- **Notable Examples**: Programs such as **FakeAV** and **Security Tool** are examples of scareware, designed to pretend to be fake security tools.
- **Impact**: Although it may not cause direct harm to the system, scareware misleads less informed users into spending money

on a fake "solution". It can lead to anxiety and loss of confidence in real security tools.

These types of ransomware represent different tactics and targets, from data encryption and system disruption to data privacy exploitation. The diversification of ransomware strategies makes it necessary to adopt a combination of security measures, such as user education, data backup and constant network monitoring, to minimize the risk of attacks and protect the digital assets of organizations and individual users.

Ryuk

Ryuk ransomware is one of the best known and most destructive examples of this class of malware. It is a type of ransomware targeting large enterprises, government institutions and hospitals, designed to cause as much damage as possible and maximize the profits obtained through extortion.

- **History and Origin**: Ryuk first appeared in 2018, linked to a cybercriminal group known as Wizard Spider, based in Russia. Ryuk has been used in targeted attacks, where victims who have the resources to pay a high ransom are carefully studied and selected.
- **Spreading Method**: Ryuk is usually spread through well-planned phishing attacks or through other ancillary malware such as **Emotet** and **TrickBot**, which function as gateways for Ryuk to infect systems. These ancillary programs allow the attacker to infiltrate the victim's network, escalate privileges and ultimately launch Ryuk on the organization's key systems.

- **Impact**: Ryuk attacks have severely affected companies in sectors such as healthcare, financial services and logistics, where downtime has serious economic and operational repercussions. For example, in 2019, several hospitals in the United States were forced to suspend services as their medical data systems were blocked by Ryuk. Ransomware figures have reached up to several million dollars, depending on the organization's ability to pay and the value of the encrypted data.

- **Extortion Strategy**: Ryuk is designed to cause such severe disruptions that affected organizations are pressured to quickly pay the ransom. Attackers use a two-pronged pressure tactic: first, they threaten to destroy the data if payment is not made; then, if the organization still does not pay, they threaten to leak the confidential information obtained.

Phishing and Ransomware Protection Strategies

Given the impact of phishing and ransomware threats, it is essential that organizations and users implement preventative measures to minimize the risk of infection and respond quickly if an attack occurs.

1. User Education and Awareness

Cybersecurity training is the first line of defense against phishing and ransomware. Organizations should train their employees to recognize suspicious emails and messages and understand the risks of clicking on unknown links or downloading unverified attachments.

Phishing simulation programs help evaluate and improve employee response to phishing emails.

2. E-mail Filtering and Network Security

Implement advanced email filters that can detect and block phishing emails before they reach users.

Use firewalls and intrusion detection and prevention systems (IDS/IPS) to monitor and analyze network traffic for suspicious attack patterns.

3. Backups and Data Recovery

Back up critical data to secure, offline locations on a regular basis. This ensures that if a ransomware attack occurs, data can be restored without the need to pay the ransom.

Periodically test data recovery procedures to ensure that backups are restored correctly and are protected from unauthorized access.

4. Anti-Phishing and Anti-Ransomware Software

Use specialized security software that detects and blocks both phishing techniques and ransomware programs. Anti-ransomware tools are capable of detecting suspicious behavior, such as mass file encryption, and blocking these activities before they spread across the network.

Regularly update all security systems and antivirus software so that they can detect and block the latest ransomware variants and other ancillary malware such as Emotet and TrickBot.

5. Implementation of Multifactor Authentication (MFA)

Use MFA to protect access to critical networks and systems. MFA adds an additional layer of security, requiring users to verify their identity through multiple factors (such as an SMS code or an authentication app), making it difficult for attackers to gain access.

6. Network Segmentation and Privilege Restriction

Implement network segmentation to limit the scope of an attack if an infection occurs. This involves dividing the network into isolated segments so that if one part of the network is compromised, the ransomware or malware cannot easily spread to other systems.

Use the principle of **least privilege**, ensuring that users only have access to the data and systems necessary for their functions. This limits the attacker's ability to escalate privileges and access critical information if they infiltrate the network.

Phishing and ransomware are increasingly sophisticated and persistent threats in today's cyber environment. Phishing attacks facilitate the introduction of ransomware such as Ryuk, a notable example of how these threats can cause significant operational disruptions and financial losses. Implementing preventative strategies, such as security training, data backup and the use of multi-factor authentication, can help significantly reduce the risk of these attacks and protect the integrity of organizations' data and systems.

DDoS Attacks and the Threat to Service Availability

What is a DDoS Attack?

Distributed *Denial of Service* or **DDoS** (*Distributed Denial of Service)* attacks seek to disable a server, website or online service by overloading its infrastructure with excessive request traffic. By consuming all the system's resources, attackers make the service inaccessible to legitimate users, causing disruptions that can be very costly for the affected companies and organizations.

These attacks typically use networks of compromised devices, known as **botnets**, which are controlled by the attackers to coordinate the mass issuance of requests. DDoS attacks can last from a few minutes to days or weeks, depending on the attackers' targets and resources.

Types of DDoS Attacks

There are several types of DDoS attacks, each aimed at exploiting specific vulnerabilities in the targeted system or network:

Volume Attacks

Volume attacks are designed to saturate the bandwidth of the target system, flooding it with a large amount of data traffic. Volume attacks can quickly consume the available bandwidth, causing services to become slow or completely inaccessible.

- **Example**: UDP *(UserDatagram Protocol)* flooding, where attackers send a large number of forged UDP packets to the server to overload the bandwidth.
- **Impact**: These attacks can affect connection speed and, in severe cases, cause network services to be completely down.

Protocol Attacks

Protocol attacks exploit weaknesses in network communication protocols, such as TCP/IP, to disrupt service. These attacks consume network resources and can crash the system by overflowing connection tables on servers and intermediate devices.

- **Example**: SYN flood attacks, where multiple SYN requests are sent (without terminating the connection) to exhaust the server's connection resources.
- **Impact**: They damage the communication protocol and render the server unable to handle legitimate requests. They are especially effective against systems without adequate security measures.

Application Layer (Layer 7) Attacks

Application layer attacks are specific attacks targeting services or applications operating at the application level (layer 7 of the OSI model). Unlike other types of attacks, application layer attacks are more difficult to detect because they simulate the behavior of legitimate users.

- **Example**: HTTP flood attack, where multiple HTTP requests are sent to the server, simulating a legitimate user access to overload the website.
- **Impact**: This type of attack can affect e-commerce sites, online banking services and other critical systems that depend on constant availability for their users.

Impact of DDoS Attacks on Organizations

DDoS attacks represent a serious threat to the availability of online services and can have a significant impact on an organization's reputation, revenue and operability. Some of the main effects include:

- **Loss of Revenue**: When a DDoS attack affects an e-commerce or financial services company, every minute of downtime represents a direct loss of revenue. For companies that rely on online services, such as e-commerce stores or gaming platforms, the disruption can be especially costly.

- **Reputational Damage**: Unavailability of a service can affect public perception and reduce trust in the company. Users expect online service organizations to provide secure and reliable access. If a website suffers frequent attacks, customers may doubt the company's ability to protect their data.

- **Recovery and Mitigation Costs**: After an attack, companies may face significant expenses to restore service, including contracting DDoS mitigation services, repairing affected infrastructure and, in some cases, paying external providers to implement preventive solutions.

DDoS Attack Case Studies

Dyn Case (2016)

In October 2016, a series of massive DDoS attacks targeting Dyn, a DNS service provider, caused worldwide disruptions to popular websites such as Twitter, Spotify, Reddit and Amazon. The attack used a botnet consisting of compromised IoT devices (such as security cameras and routers) via the **Mirai** malware.

- **Method**: Mirai scanned the internet for vulnerable IoT devices using factory credentials and infected them, forming a botnet. The botnet then launched a volume attack that affected Dyn, causing DNS to fail and rendering sites around the world inoperable.
- **Impact**: This attack highlighted the vulnerability of IoT devices and their potential to be used in DDoS attacks. It was an important reminder of the need for security on connected devices.

GitHub Case (2018)

In 2018, the collaborative development site GitHub suffered the largest DDoS attack on record to that date, with traffic volume reaching 1.35 Tbps. The attack employed a technique called **memcached amplification**, in which attackers exploited misconfigured memcached servers to increase the volume of the attack.

- **Method**: The attackers sent forged requests to the memcached servers, which in turn responded with large volumes of data to the GitHub website, amplifying the amount of traffic received.
- **Impact**: GitHub was able to mitigate the attack within minutes thanks to its defense infrastructure, but the attack demonstrated the ability of cybercriminals to launch massive attacks in a short time.

Defense Strategies against DDoS Attacks

To protect against DDoS attacks, organizations must adopt a combination of preventive measures and mitigation technologies to manage malicious traffic and minimize the impact on legitimate services.

Implementation of Firewalls and Intrusion Detection Systems (IDS)

Firewalls and IDS systems can monitor and filter network traffic for DDoS attack patterns. While firewalls will not stop a large-scale attack, they help reduce malicious traffic in the early stages of the attack.

DDoS Mitigation Services

Many companies choose to engage specialized DDoS mitigation services, which have advanced infrastructure to detect and divert attack traffic before it reaches the organization's network. These services can include cloud-based solutions that automatically scale to handle large volumes of malicious traffic.

Content Delivery Networks (CDN)

CDNs distribute a website's content across multiple servers around the world, reducing the load on a single server and improving responsiveness in the event of an attack. CDNs can also provide DDoS mitigation capabilities by redirecting traffic and limiting the impact on the organization's core infrastructure.

Load Balancing

The use of **load balancers** allows traffic to be distributed among several servers to prevent a single server from being overwhelmed. Load balancing is an effective technique for handling traffic peaks and reducing the risk of infrastructure collapse.

Cloud Infrastructure Scalability and Resiliency

Cloud platforms offer the ability to scale the infrastructure in real time to adapt to the volume of traffic. This allows systems to absorb traffic spikes caused by a DDoS attack without interrupting service.

Implementation of Rate Limiting Policies

Limit rate policies can restrict the number of requests a user can make in a given period. This limits the impact of application layer attacks, such as HTTP flood attacks, by preventing a single user or IP address from sending a large number of requests.

DDoS attacks represent a constant threat to the availability of online services and can have a devastating impact on organizations, both in financial and reputational terms. Defense strategies and DDoS mitigation technologies are essential to ensure that systems remain operational, even in attack situations. Implementing solutions such as CDNs, cloud mitigation services and firewalls helps minimize the risk of outages and protect assets.

Chapter 3: Defense Strategies in Cybersecurity

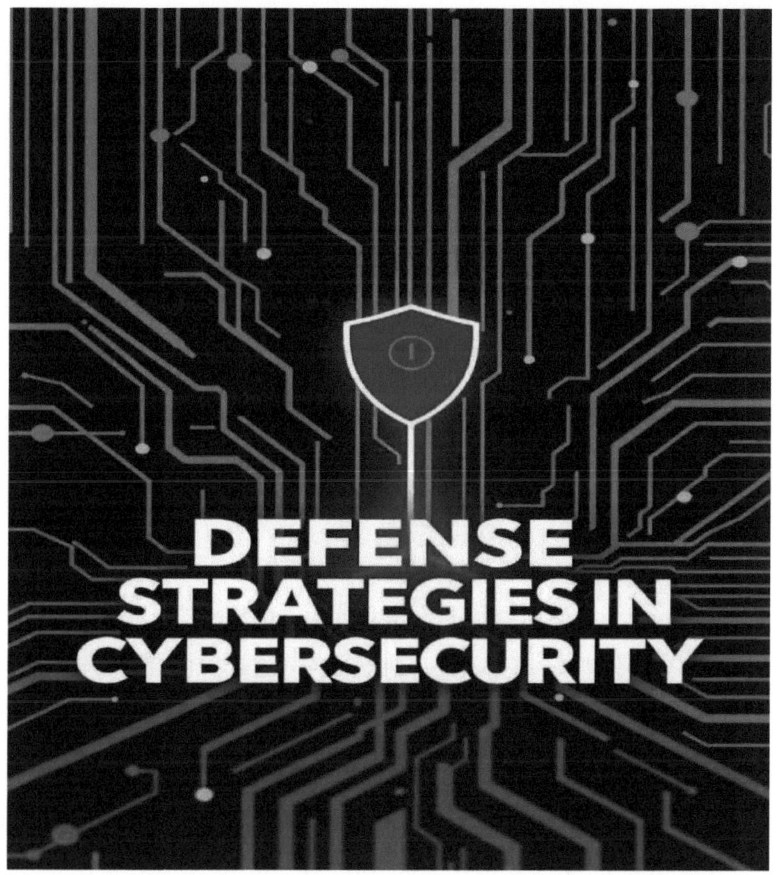

Multifactor Authentication and Encryption: The First Line of Defense

Multi-factor authentication (MFA) has become a standard for ensuring that access to critical accounts and systems is properly protected. Unlike traditional authentication, which relies only on a **password**, MFA requires the user to provide more than one authentication factor. This makes access significantly more difficult for attackers, even if they manage to obtain one of the user's credentials. Commonly used factors include:

- **Something the user knows**: a password or PIN. Although it is the most common factor, it is not enough on its own due to its vulnerability to attacks such as **phishing** or **brute force**.
- **Something the user has**: a security token, smart card or authentication application (such as Google Authenticator or Authy). These applications generate temporary codes that must be entered along with the password, which adds an extra layer of security.
- **Something the user is**: biometrics, such as fingerprints, facial recognition or iris identification. These factors are becoming increasingly common, especially in mobile devices and high-security systems.

MFA is a crucial defense against many types of attacks, especially those that focus on **phishing**. Although a user's password can be easily obtained through **phishing** or **keylogging** techniques, an attacker who does not possess the second factor, such as the user's cell phone, will not be able to complete access.

Additional benefits of MFA include:

- **Reduced risk of unauthorized access**: Even if an attacker obtains the password, he would need the second authentication factor.
- **Increased protection for critical services**: MFA is especially important for platforms such as banks, email services and social networks, where unauthorized access can have serious consequences.
- **Regulatory compliance**: Many regulations and security standards, such as **PCI-DSS** and **GDPR**, require the implementation of MFA to protect users' personal and financial information.

On the other hand, **encryption** is a fundamental process that guarantees the confidentiality and integrity of data, even if they are intercepted. Encryption applies to both data **in transit** (when traveling over networks) and data **at rest** (when stored on disks, databases, etc.). There are several types of encryption algorithms, but the most common and robust include:

- **AES (Advanced Encryption Standard)**: This symmetric encryption algorithm is widely used and is considered highly secure. In AES, the same key is used to encrypt and decrypt data, and its variant AES-256 is one of the most recommended due to its security level.
- **RSA (Rivest-Shamir-Adleman)**: An asymmetric encryption algorithm that uses a public and private key pair. The public key is used to encrypt data, while the private key is used to decrypt

it. It is commonly used in applications such as **SSL/TLS** to protect web communication.

- **TLS/SSL (Transport Layer Security/Secure Sockets Layer)**: Although SSL is no longer considered completely secure, TLS is still the technology used to secure communications over the web. TLS encrypts the connection between the server and the client, preventing attackers from intercepting or modifying the transmitted information.

Encryption not only guarantees the **confidentiality** of information, but also protects against data **integrity**, preventing it from being altered without detection. To ensure an adequate level of protection, it is essential to use up-to-date encryption protocols and algorithms, and also to ensure that encryption keys are securely managed.

Importance of encryption key management:

- **Rotation and secure storage**: Encryption keys should be rotated regularly and stored securely, using tools such as **HSMs (Hardware Security Modules)**, which allow for more robust key management.

- **Backup and recovery**: It is crucial to have key backup mechanisms and disaster recovery policies in place, as losing encryption keys can result in irreversible data loss.

Firewalls, IDS and IPS: Monitoring and Prevention Tools

Monitoring and prevention tools, such as **firewalls**, **Intrusion Detection Systems (IDS)** and **Intrusion Prevention Systems (IPS),** are critical components in any cyber defense strategy. These technologies work together to protect network infrastructure and systems, ensuring that legitimate traffic flows smoothly, while blocking malicious access and detecting suspicious activity.

Firewalls

Firewalls act as a **filter** between internal and external networks, controlling incoming and outgoing traffic. They can be **hardware** or **software**, and their main function is to apply predefined rules to allow or deny connections according to the type of traffic, IP address, port, protocol and other factors.

There are different types of firewalls, among them:

- **Packet Filtering Firewalls (Packet Filtering Firewalls)**: These firewalls inspect data packets passing through the network and filter them according to predetermined rules. If the packet meets the set criteria (e.g. IP address, port or protocol), access is allowed; otherwise, it is blocked.
- **Deep Packet Inspection (DPI) firewalls**: These are more advanced than traditional firewalls, as they not only check the headers of data packets, but also the contents of the packet for threat patterns or viruses. This makes it possible to detect more sophisticated attacks.
- **Next-Generation Firewalls (NGFW)**: These are an evolution of traditional firewalls and include advanced capabilities such as encrypted traffic inspection, threat intelligence integration, and the ability to identify and block malicious applications and behavior, such as **C2 (Command and Control)** botnets.

Firewalls help prevent unauthorized access and are essential in creating a **demilitarized zone (DMZ)** in the network infrastructure, separating public networks (such as the Internet) from internal networks (such as the corporate intranet).

Intrusion Detection Systems (IDS)

An **IDS** is a system that monitors network traffic or computer systems for **suspicious behavior** that may indicate an attempted attack or intrusion. The goal of an IDS is to **detect** attacks in real time, but not to block them automatically. Instead, an IDS generates **alerts** so that security administrators can review and take corrective action.

There are two main types of IDS:

- **Network-based IDS (NIDS - Network IDS)**: Monitors traffic flowing through the network, looking for patterns or anomalies that match known attack signatures (such as port scans or bot traffic).
- **Host-based IDS (HIDS)**: Monitors specific events within a host or device, such as operating system logs, file access and software changes. It is effective in detecting internal attacks or those targeting specific vulnerabilities on a device.

IDSs can identify a wide range of threats, such as **port scans**, vulnerability **exploits**, **privilege escalation attempts**, among others. However, their ability to prevent intrusions is limited, as they do not have the capacity to block traffic automatically. Their main function is the **early detection** of threats.

Intrusion Prevention Systems (IPS)

IPSs are similar to IDSs, but with a key difference: they **not only detect** attacks, but actively **prevent** malicious access by blocking traffic **in real time**. When the IPS detects a threat, it interrupts the attacker's connection or blocks the malicious packet before it can do damage to the network.

IPS can act in different ways:

- **Signature IPS**: Uses a database of known attack signatures and compares traffic against these signatures. If the traffic matches an attack signature, the IPS blocks the suspicious traffic.
- **Anomaly-based IPS**: Instead of looking for known signatures, the IPS analyzes network behavior and creates a normal traffic profile. If it detects a significant deviation from this pattern, it classifies it as a potential attack and blocks the traffic.
- **Hybrid IPS**: Combines features of both approaches, using both signature detection and anomaly analysis to improve threat detection and prevention.

Importance and Key Differences

- **Firewalls**: They provide a basic barrier that allows legitimate traffic and blocks unauthorized traffic. They are the first line of defense and must be well configured.
- **IDS**: They focus on detecting suspicious behavior and alert

administrators, but do not automatically prevent attacks. They are essential for detecting **zero-day** attacks or new threats.

- **IPS**: They offer a more advanced layer of protection, as they not only detect threats, but also act to prevent threats from affecting the system or network.

In summary, while **firewalls** protect networks and **IDS/IPS** focus on intrusion detection and prevention, together they form a comprehensive defense system that provides constant monitoring, prevention of unauthorized access and the ability to detect and mitigate attacks in real time.

Awareness and Training: Preparing Personnel to Defend Themselves

Cybersecurity awareness and **training** are critical components of a comprehensive security strategy. While advanced technologies such as firewalls, IDS, IPS, and multi-factor authentication are essential to protect systems, the weakest link in the defense chain remains **the human being**. Attackers often exploit human vulnerabilities through tactics such as **phishing**, **social engineering**, and **operational errors**. This is why **continuous education** and a **security-focused organizational culture** are crucial to mitigate risks.

Cybersecurity awareness

Cybersecurity awareness involves making employees aware of the security threats and practices they need to adopt on a daily basis. An effective awareness program has several objectives, including:

- **Recognize threats**: Employees must be able to identify different types of cyber-attacks, such as **phishing**, **malware**, **ransomware**, **DDoS attacks**, and how they can protect themselves against them.

    - **Phishing**: Training should include examples of fraudulent emails and how to avoid them, such as verifying the authenticity of the sender, not clicking on suspicious links and not downloading attachments from untrusted sources.

    - **Ransomware**: Explain how this type of attack can come via email, malicious websites or infected links, and how to protect yourself by making regular backups and updating your software.

- **Best practices**: Teach basic security practices that employees should follow in their professional and personal lives, such as using **strong passwords**, **changing passwords regularly**, and **multi-factor authentication**.
    - **Strong passwords**: Employees must understand the importance of not reusing passwords and of using combinations of letters, numbers and special characters.

    - **Multi-factor authentication (MFA)**: Reinforce that enabling MFA adds an extra layer of security, especially in critical services.

- **Security policies and regulations**: Employees should be familiar with the organization's security policies, such as rules on the use of personal devices on the corporate network

(BYOD), classification and protection of sensitive information, and guidelines for the use of public Wi-Fi networks.

Cybersecurity Training

Cybersecurity training goes beyond awareness and provides employees with the technical knowledge and skills needed to effectively manage threats. This includes both basic training for all employees and advanced training for specific roles, such as system administrators and IT staff.

- **Basic training**: In this training, employees learn the fundamental concepts of cybersecurity, how to identify fraudulent emails, how to handle confidential information and how to respond to incidents. In addition, the use of basic security tools, such as antivirus and **personal firewalls**, should be taught.
- **Advanced training**: This type of training is aimed at IT personnel and cybersecurity professionals. It includes more technical topics such as **firewall configuration**, **incident management**, cyberattack forensics and critical infrastructure protection. It is also important to teach about **cryptography**, **network monitoring** and the use of **intrusion detection and prevention systems (IDS/IPS)**.
- **Attack simulations**: Performing practical **attack simulation** exercises such as phishing attacks, **mock security incidents** or penetration tests (pen tests) allows employees to experience

in real time how to react to a cyber-attack. This also helps to improve responsiveness and identify weaknesses in defense systems.

- **Incident management**: Training staff on how to respond to a security incident is essential. This includes identifying the attack, reporting it, taking immediate action to mitigate damage, and coordinating the response with other departments and experts. An **incident response plan** should be implemented that details the procedures to follow in the event of a security breach.

Organizational safety culture

Fostering an **organizational culture of Cybersecurity** means integrating security into all aspects of the business, from decision making to day-to-day operations. This includes:

- **Senior management commitment**: Organizational leaders must lead by example on the importance of cybersecurity and demonstrate an active commitment to protecting digital assets. Security must be integrated into the company's goals and policies.
- **Positive reinforcement**: Rewarding and recognizing employees who follow good cybersecurity practices can motivate others to do the same. This may include incentives or formal recognition for those who demonstrate exemplary implementation of security practices.

- **Continuous updating**: As cyber threats evolve rapidly, **continuous** training and awareness and regular updates on new threats, technologies and security protocols are essential. In addition, additional training should be provided to employees whenever new security tools or processes are implemented.

Benefits of Awareness and Training

1. **Reduced security incidents**: By properly training staff, the likelihood of employees falling into social engineering traps, such as phishing, is reduced.
2. **Fast and effective responses**: Well-trained personnel can quickly identify, mitigate and report threats, minimizing the impact of incidents.
3. **Regulatory compliance**: Security training also helps organizations comply with regulations and security standards, such as **GDPR** or **PCI-DSS**, which require employees to receive data protection training.

Cybersecurity **awareness and training** are key pillars of any cyber defense strategy. By training staff and fostering an organizational culture of security, companies not only protect their digital assets, but also empower their employees to act as a **first line of defense** against cyber threats.

Chapter 4: Incident Management and Cyber Attack Response

Early Detection: Identifying the Signs of an Incident

Early detection is one of the most important pillars of an effective response to cyber-attacks. Detecting an incident as quickly as possible allows minimizing the damage, containing the threat and activating response plans efficiently. To achieve effective early detection, it is essential to have an adequate monitoring system, as well as the ability to identify unusual patterns or behaviors in systems and networks.

Continuous Infrastructure Monitoring

Continuous monitoring is essential to detect suspicious activity before it evolves into a larger scale incident. Systems must be configured to provide full visibility into infrastructure components and applications.

- **Network Monitoring Systems (NMS)**: These tools allow observing network traffic in real time, identifying unusual increases in data volume or anomalous patterns in communication between devices. For example, in a **DDoS** attack, the system could detect a drastic increase in incoming traffic to a server, indicating an attempt to saturate the network.
- **Application Monitoring Systems**: Applications also need to be monitored for anomalous behavior, such as unusual access requests or malicious user behavior. **Intrusion Detection Systems (IDS)** can identify patterns that suggest an attacker is exploring vulnerabilities within a web application, such as a **SQL Injection** attack.
- **File and Log Integrity Monitoring**: Tools that monitor **file**

integrity can detect unauthorized changes or the addition of malicious files. System **logs** (event logs) are critical for early detection of an attack, as every action within a system generates detailed logs. Monitoring these logs can indicate unauthorized activity, such as access attempts or modification of important files.

Logs and Event Data Analysis

Logs generated by servers, applications and network devices are an invaluable source for early detection. Analysis of these logs can help identify unusual behavior that could be indicative of an attack.

- **Real-Time Log Analysis**: The use of tools such as **SIEM (Security Information and Event Management)** allows the collection and correlation of logs in real time. SIEM solutions can identify anomalous patterns, such as access to files or systems outside normal hours, multiple failed login attempts or unauthorized changes to critical configurations.
- **Security Log Analysis**: Logs from **firewalls**, **IDS/IPS** and **anti-virus systems** provide key information about possible intrusion attempts, detected malicious traffic and other signs of suspicious activity. Logs can also contain details about the location, type of attack and resources compromised, which is crucial for rapid response.

Indicators of Commitment (IoCs)

Indicators of Compromise (IoCs) are artifacts that can signal that a network or system has been compromised. These can include malicious IP addresses, malware signatures, corrupted file hashes or malicious domains. Having an up-to-date IoC database and monitoring these indicators in the infrastructure helps detect security compromises quickly.

- **Common IoCs**: Examples include:
  - **IP addresses** of command and control servers used by attackers.
  - **URLs or domains** related to the malware or the attacker's activities.
  - **Malicious file hashes** that match known malware.
  - **Signals of anomalous execution** of unknown scripts or processes.

Implementing systems that can automatically search network **logs** or IoC databases can significantly improve the ability to detect incidents in a timely manner.

Anomalous User Behavior

Many cyberattacks, such as **phishing**, **spoofing** or **lateral movement** (when an attacker moves within a network after compromising a system), often begin with legitimate access by a compromised user. Monitoring user behavior can provide early signs that a system has been infiltrated.

- **User Behavior Analytics (UBA)**: **User and Entity Behavior Analytics (UEBA)** tools can identify anomalous patterns in user activity, such as an unusual volume of downloads, access to unauthorized resources, or logins from unusual geographic locations.
- **Authentication and Access**: An increase in failed login attempts, use of credentials from unrecognized devices or unusual locations, or access to data after hours are signs that should be closely monitored. These behaviors may indicate a **brute force attack**, a **phishing** attempt, or the presence of **malware** that has compromised credentials.

Malware and Malicious Software Indicators

Malware attacks can take many forms, from simple viruses to sophisticated **ransomware**. Identifying the presence of malware early is critical to mitigating the effects of an attack.

- **Malware Signature**: Antivirus solutions and malware detection tools can identify malicious files on systems and their characteristic signatures. Files or processes that do not match the signature of legitimate software should be investigated immediately.
- **Malware behavior**: **Sandboxing** and dynamic analysis tools can help detect malware behavior, such as communicating with command and control servers, or modifying files and system settings. Malware programs may attempt to hide, so monitoring

processes and changes to critical files is critical.

Advanced Tools for Early Detection

To improve early detection, organizations can use advanced cybersecurity tools that integrate multiple data sources and automated analytics to identify suspicious patterns more quickly.

- **Artificial Intelligence (AI) and Machine Learning (ML)**: AI and ML-based solutions are designed to learn normal patterns in network traffic and system activities. By analyzing large amounts of data, these tools can identify anomalous behavior or new tactics used by attackers, even if they have not been previously documented.
- **Predictive Analytics**: Predictive analytics can help predict potential attack vectors based on historical trends and current data. This can enable organizations to anticipate attacks and take preventative measures before they materialize.

Early detection depends on constant monitoring, effective log analysis, the use of threat intelligence tools, and the ability to identify anomalous patterns in system or user behavior. Having a robust **monitoring and analysis** infrastructure in place helps to identify potential attacks in their early stages and take appropriate measures to mitigate risks. If implemented correctly, early detection can significantly reduce the impact of a cyber-attack and facilitate a quick and efficient response.

Response Plan: How to Act During a Cyber Attack?

When an organization faces a cyber attack, it is essential to have a well-defined **response plan** to mitigate damage and restore operations as quickly as possible. The lack of a structured plan can result in a disorganized reaction, increasing the duration of the attack and the impact on the company's digital assets. A response plan should be clear, scalable and aligned with the organization's security objectives.

Preparedness and Planning: The Foundation of Effective Response

Before a cyber-attack occurs, the organization should have a well-structured incident response plan. This plan should define roles, responsibilities and detailed procedures for each possible type of incident. The goal is to be fully prepared to act quickly and efficiently.

- **Definition of Roles and Responsibilities**: An **incident response team** (IRT) should be designated in advance. This team will be composed of cybersecurity, IT, legal, communications and management experts, and each member should have clearly assigned tasks and responsibilities. This includes personnel to handle internal and external communications, personnel to manage technology solutions, and others to manage legal or compliance issues.
- **Clear Communication Procedures**: The plan should establish

how the organization will communicate during and after the attack. This includes both **internal communication** (to inform employees about the attack and next steps) and **external communication** (with customers, partners, and regulators). Transparency is key to maintaining stakeholder confidence during an attack.

- **Drills and Ongoing Training**: Regular staff training is crucial. Conducting **incident response drills** allows the team to become familiar with the process and identify areas for improvement. These exercises also help all team members understand their specific role during the incident.

Incident Detection: First Steps in Detecting an Attack

Early detection of an incident is critical, but so is the appropriate initial response. Once an attack is detected, the **response plan** must be activated. The first steps are crucial to contain the damage and prevent escalation of the attack.

- **Incident Confirmation**: Sometimes alerts can be false, so it is necessary to verify if it is really an incident. This can be done by reviewing system logs, verifying **IDS/IPS** alerts or by analyzing alerts of anomalous network traffic. An attack may include indicators such as unauthorized access to systems or applications, unexpected changes to files or configurations, or unusual network activity.
- **Response Plan Activation**: Once confirmed as an incident, the

response plan should be activated and all members of the response team should be notified immediately. The plan should provide for how to escalate the incident according to its severity (from a small attack to a large-scale attack affecting critical infrastructure).

Containment: Limiting the Scope of Attack

Containment seeks to isolate the attack to prevent it from spreading or causing further damage. Depending on the type of attack, containment measures may vary, but the common goal is to mitigate the impact on the infrastructure.

- **Isolate Affected Machines or Networks**: If the attack has affected a specific system, such as a server or workstation, it may be necessary to **disconnect** this system from the network to prevent the attack from spreading to other devices or systems. In the case of **ransomware** attacks, the malware must be prevented from spreading to other machines through the shared network.
- **Applying Firewall Rules**: During an attack, it can be useful to apply specific firewall rules to block malicious traffic or traffic from certain known malicious sources (IP addresses, ports, protocols). This helps to limit communication between compromised systems and attackers, which could reduce damage.
- **Blocking Compromised User Accounts**: If the attack involves

the use of compromised credentials, the user accounts being exploited by the attackers should be blocked immediately. This may involve resetting passwords, applying **multi-factor authentication (MFA)**, or temporarily disabling compromised accounts.

Eradication: Eliminate the Threat

Eradication is the complete removal of the threat from the system. This is a crucial step to ensure that the attack cannot be restarted or affect the infrastructure again.

- **Malware Removal**: If **malware** has been identified as having infiltrated the system, the response team should proceed to remove any malicious code or infected components. This may include deleting files, cleaning malware logs, and using antivirus or **sandboxing** tools to ensure the system is clean.
- **Vulnerability Remediation**: **Vulnerabilities** that attackers exploited to compromise the system must be identified. This includes fixing missing security patches, improving security configurations or updating compromised passwords. Ensuring that the infrastructure has no backdoors that attackers could have left behind is essential.

Recovery: Back to Normal

After eradication of the threat, the goal is to **restore operations** as quickly as possible, while ensuring that systems are secure.

- **System and Data Restoration**: Systems should be restored from the most recent **backups** that are known to be uncompromised. It is important to validate that the restored systems are free of malware and that there is no persistence of the threat. Databases, files and applications should be carefully reviewed before being restored to production.
- **Functionality Testing**: Once systems are restored, they must be thoroughly tested to ensure that applications, services and networks are back to operating as expected with no additional bugs or vulnerabilities.

Lessons Learned: Improving Defense for the Future

Once the incident has been contained and operations have been restored, it is critical to conduct a **post-incident review** to learn from the experience and improve defenses in the future.

- **Forensic Analysis**: A detailed analysis of the incident should be performed to understand how the attack occurred, what techniques the attackers used and what weaknesses in the systems allowed unauthorized access. This forensic analysis helps identify important lessons and areas for improvement.
- **Response Plan Review and Update**: As lessons are learned from incidents, the response plan should be updated. Detection, containment, eradication and recovery procedures should be improved based on experience and new emerging threats.
- **Ongoing Training**: Based on lessons learned, organizations

should improve training for personnel and conduct more drills. This will help the response team be more prepared for future threats.

Post-Incident Communication

An essential part of responding to a cyber-attack is **communication management**. It is critical to communicate effectively both internally and externally.

- **Internally**, employees should be informed about the incident and the actions they should take (e.g., change passwords, avoid certain websites). In addition, **managers and executives** should have access to detailed information about the impact and actions taken.
- **Externally**, **customers** and **partners** should be notified in a timely manner, especially if the attack compromises sensitive data. It is important to be transparent without revealing details that could compromise the security of the organization.

The **incident response plan** should be viewed as a continuous process of preparation, detection, response, and improvement. Having a well-defined strategy and a skilled team can make the difference between an effective response that minimizes damage and a disorganized reaction that exacerbates the situation. An organization that has clear protocols and is able to adapt quickly during a cyberattack will be much better positioned to reduce the impact of the incident and learn from each experience.

Recovery and Continuous Improvement: Learning from Incidents

Recovery from a cyber-attack is only the first step in the process of restoring normalcy within the organization. However, the real key to protecting against future incidents lies in **continuous improvement** based on lessons learned from the incident. Recovery is not just about restoring services and systems, but ensuring that the same mistakes are not repeated and that the organization's defenses become increasingly robust.

System and Service Restoration: Back to Normalcy

After containing and eradicating the attack, **recovery** becomes an immediate priority. This involves restoring the systems and services affected by the incident to ensure the organization's operational continuity.

- **Restoration of Affected Systems and Data**: In the recovery phase, the compromised systems, servers, applications and databases must be restored from pre-attack **backups**. It is essential to ensure that the backups are up-to-date and reliable. It is also necessary to perform extensive testing to verify that the restored systems are fully functional and free of malware.
- **System Integrity Validation**: Before allowing users to access the restored systems, security checks should be performed to ensure that there are no backdoors, vulnerabilities or persistent malware in the infrastructure.
- **Testing the Effectiveness of Containment and Eradication**

Solutions: During recovery, it is important to validate that the containment and eradication measures taken during the attack were effective and that the attackers no longer have access to the system.

Lessons Learned: Post-Incident Analysis and Reflection

Once the environment has been restored, it is essential to conduct a **post-incident analysis** process to understand how the attack occurred and what could have been done better. This analysis not only helps to detect the vulnerabilities that were exploited, but also to strengthen the organization's defense.

- **Forensic Analysis of the Incident**: Forensic analysis involves investigating in detail how the attack was carried out, identifying which systems were compromised, what attack methods were used, and what the entry vector was. This investigation helps determine if there were gaps in the defenses that allowed the attackers to gain access.
- **Root Cause Identification**: Beyond the actions of the attacker, it is essential to understand the **root causes** of the incident. Was it a software vulnerability? Was it human error? Was there a failure in monitoring procedures or internal security practices? Identifying the root causes allows you to implement long-term solutions that will prevent the same type of attack from happening again.
- **Team Response Review**: Evaluating how the response team handled the incident is key to improving organizational capabilities. This includes reviewing reaction times, internal

communication, decision making and the measures taken to contain and eradicate the threat.

Improving Defenses: Strengthening Security

Based on lessons learned and post-incident analysis, the organization must take steps to **strengthen its defenses** and reduce the likelihood of a similar attack occurring again. This involves improving both **security policies** and **protection tools**.

- **Updating Security Policies**: If the incident revealed flaws in security policies, these should be reviewed and updated. For example, access control, authentication, or encryption policies may need to be tightened. It is critical to integrate any policy changes into the organization's overall **risk management** framework.
- **Strengthening Defense Tools**: If the attack exploited a vulnerability in a security tool, such as a firewall or intrusion detection system (IDS), its upgrade or replacement should be considered. In addition, new security technologies, such as **artificial intelligence** and **machine learning**, should be integrated to improve detection and response to future threats.
- **Implementing New Layers of Security**: **Defense in depth** is essential. Adding new layers of security, such as **multi-factor authentication (MFA)**, **end-to-end encryption**, and emerging technologies such as **Zero Trust** can help prevent future attacks, even if an attacker manages to penetrate a layer of defense.

Training and Awareness: Training Personnel

One of the most valuable lessons from any incident is the need to **train personnel** and improve **security awareness**. Many times, attacks take advantage of **human interaction**, as in the case of **phishing** or system configuration errors. Therefore, it is critical to implement an ongoing cybersecurity awareness and training program.

- **Ongoing Training Programs**: Ensuring that all employees receive regular training on security best practices and how to detect potential threats is critical. Training should include how to handle passwords securely, how to identify suspicious emails and what to do if suspicious activity is detected.

- **Cybersecurity Drills**: **cyber-attack drills** allow you to test your staff's knowledge in realistic scenarios. This includes phishing simulations, penetration testing and other attack scenarios. These exercises help employees improve their ability to recognize and respond to cyber threats.
- **Organizational Security Culture**: Creating a **security culture** within the organization is key. This involves promoting the idea that cybersecurity is everyone's responsibility, from senior management to rank-and-file employees. Fostering a culture of alertness and shared responsibility increases the likelihood of detecting an attack before it causes significant damage.

Review and update of the Incident Response Plan.

Cyberattacks and threats evolve rapidly, so the **incident response plan** must be reviewed and updated after each event. The organization must be prepared to respond to new types of threats or more sophisticated attacks.

- **Post-Incident Evaluation of the Response Plan**: After each attack, the response plan should be evaluated to determine its effectiveness. Were the established procedures followed? Were there delays? Were the security tools adequate? These questions should guide continuous improvement of the plan.
- **Adjustment of Procedures and Tools**: Based on the incident analysis and lessons learned, response procedures and defense tools should be adjusted. This could involve adopting new technologies, updating communication protocols, or revising roles within the response team.

Stakeholder Report: Transparency and Confidence

Once the incident has been handled and operations have been restored, it is vital to **inform stakeholders in** a transparent manner about what happened. This includes both **internal** (employees) and **external** (customers, suppliers, partners and regulators) **communication**.

- **Post-Incident Report**: The report should detail what happened, how the incident was handled, what steps were taken to mitigate the damage, and what changes are being implemented to prevent future incidents. **Transparency** in post-incident communication is critical to maintaining stakeholder confidence.

- **Regulatory Compliance**: Depending on the nature of the attack, it may be necessary to comply with **legal and regulatory requirements**, such as notifying data breaches to data protection authorities. Organizations should ensure compliance with local and international laws in case of incidents involving sensitive data.

Recovery and **continuous improvement** are essential processes that enable organizations to not only overcome cyber-attacks, but also strengthen their long-term security. Learning from incidents, implementing improvements to defenses, training staff and updating response plans are critical steps to ensure that the organization is better prepared for future cyber challenges. Cybersecurity is an ongoing process that requires adaptability, proactivity and a strategic approach.

Chapter 5: The Future of Cybersecurity: Trends and New Technologies

Cybersecurity is rapidly evolving to address increasingly complex and sophisticated threats in an interconnected digital world. This chapter focuses on emerging technologies that are transforming the way organizations protect their most valuable assets. As cyberattacks become more sophisticated, traditional solutions are no longer sufficient. In this context, **artificial intelligence (AI)**, **machine learning (ML)**, **proactive Cyber Defense** and **Cloud Cybersecurity** are emerging as key enablers for the future of cyber protection.

Artificial Intelligence and Machine Learning in Cyber Protection

Artificial Intelligence (AI) and **Machine Learning (ML)** have emerged as key tools in defending against cyber threats. Both technologies are revolutionizing the way organizations detect, prevent and respond to attacks, bringing advanced capabilities to handle the increasing complexity of modern cyberattacks.

Introduction to AI and ML in Cybersecurity

- **Artificial Intelligence (AI)**: Refers to the simulation of human intelligence processes by machines, especially computer systems. AI in cybersecurity is used to perform analysis and decision-making tasks that previously required human intervention. AI can learn patterns, detect anomalies, foresee threats and make autonomous decisions based on the data it receives.

- **Machine Learning (ML)**: A subfield of AI that relies on the ability of machines to learn from data without being explicitly programmed. Through algorithms, ML allows systems to adapt and improve over time as they process more information and face new threats.

The two technologies complement each other, as ML enables AI systems to continuously improve their performance without the need for constant intervention, increasing the effectiveness of cyber defenses.

Key Applications of AI and ML in Cybersecurity

Real-Time Threat and Anomaly Detection

One of the main applications of AI and ML is threat detection. Traditional systems often struggle to identify patterns of anomalous behavior due to their reliance on predefined rules. However, AI and ML can analyze large volumes of data and detect unusual or suspicious behavior in real time, such as unusual network activity, unauthorized connections, or traffic patterns that suggest a DDoS attack.

- **Predictive Analytics Algorithms**: By using ML algorithms, systems can predict attacks before they occur, based on historical patterns and by analyzing early warning signs that might go unnoticed by traditional methods. This includes identifying vulnerabilities in applications, infrastructure and user behavior.
- **Intrusion Detection (IDS)**: Traditional intrusion detection systems base their alerts on known attack signatures. However,

ML enables the creation of **adaptive IDS** systems that can identify new or unknown attack patterns, adjusting their behavior as they learn.

Malware and Ransomware Identification

Malware and ransomware are some of the most common types of threats, and attacks using them can be difficult to detect due to their ability to evolve and camouflage themselves. AI and ML-based systems can detect these threats more accurately than traditional approaches by analyzing files and software behaviors for patterns that suggest malicious activity.

- **Malware Classification**: ML models can classify files according to their behavior and characteristics, quickly identifying files that could contain malware. In addition, these models are able to detect previously unknown (zero-day) malware by analyzing their characteristics and behavior patterns.
- **Ransomware**: Ransomware uses sophisticated techniques to encrypt files and extort money from victims. AI and ML systems can detect these attacks by analyzing sudden changes to files or the network, and then trigger containment measures before the attack is complete.

Automatic Prevention and Incident Response

AI and ML not only help identify and analyze threats, but can also automate incident responses. When a threat is detected, systems can execute response actions without human intervention, such as blocking malicious IP addresses, isolating compromised machines or stopping certain processes that could be involved in an attack.

- **Incident Containment Automation**: If a threat is detected, AI systems can make automated decisions to contain the attack, such as disconnecting a server from the network or quarantining suspicious files, thereby reducing exposure time and potential damage.

- **Adaptive Responses**: AI systems can learn from previous incidents and adjust their responses as they face new types of threats. This means continuous improvement in response effectiveness and greater efficiency in cyber protection.

False Positive Reduction

A significant challenge in cybersecurity is the high rate of **false positives**, where security systems generate erroneous alerts that can lead to work overload for security teams. AI and ML-based systems have the ability to significantly reduce these false positives by learning to distinguish between real threats and benign events.

- **Intelligent Filtering**: As systems process more data, ML algorithms improve their ability to filter alerts, increasing accuracy and reducing the number of false occurrences that need to be investigated.

Identity and Access Management (IAM)

Identity and access management (IAM) is a critical component in cybersecurity, and AI and ML solutions improve organizations' ability to manage who has access to what resources, and how to detect unauthorized access. AI algorithms can analyze login patterns and behavior to identify users who are acting suspiciously, even if they have not violated specific rules.

- **Adaptive Authentication**: AI systems can apply access policies based on user behavior and context, enabling adaptive authentication. This means that the system can request a second verification if it detects something unusual, such as access from an unusual location or device.

Challenges and Considerations in the Use of AI and ML in Cybersecurity

While AI and ML offer powerful solutions, there are also significant challenges and considerations:

- **Lack of Transparency (Black Box)**: AI and ML algorithms can be complex and opaque, making it difficult to understand how they make decisions. This can be a problem if security systems make recommendations or take actions without security analysts being able to easily interpret the reasoning behind them.

- **Model Training**: For AI and ML-based systems to be effective, they must be trained on large volumes of high-quality data. This can be challenging, as the data must be representative of real threats, and the training process can be costly and time consuming.

- **Evolving Attacks**: As cybercriminals also adopt AI and ML in their attacks, defense systems must continue to evolve to keep up with new tactics and techniques.

The Future of AI and ML in Cybersecurity

As AI and ML continue to advance, we are likely to see even greater integration of these technologies into cybersecurity solutions. The future of cyber protection will be marked by increasingly autonomous systems that are able to continuously learn and adapt to threats, providing proactive and adaptive defense against attacks.

- **Predictive Cybersecurity**: AI and ML systems will not only detect threats, but also predict future attacks, helping organizations strengthen their defenses before they occur.
- **Human-Machine Collaboration**: Artificial intelligence will not replace cybersecurity experts, but will support them by providing advanced tools that allow them to focus on more strategic and less repetitive tasks.

Artificial intelligence and **machine learning** are transforming the way organizations approach cybersecurity. These technologies enable faster and more accurate threat detection, automated incident responses, reduced false positives and improved identity management, making them essential elements in the defense against

Proactive Cyber Defense: Beyond Reaction to Threats

Proactive cyber defense refers to a preventative approach in which organizations seek to anticipate, mitigate and prevent cyber attacks before they occur. Rather than relying solely on reactive response, where action is taken after a threat has materialized, proactive cyber defense seeks to identify and neutralize threats in their early stages, or even before they are launched.

This approach has become increasingly essential due to the sophistication and frequency of cyber-attacks. Attackers are using increasingly advanced tools and techniques, so relying only on detecting threats after they have occurred is insufficient. Proactive cyber defense, therefore, is key to minimizing risks and reducing potential damage.

Main Components of Proactive Cyber Defense

Vulnerability Assessment and Risk Analysis

One of the first actions in a proactive cyber defense strategy is **continuous vulnerability assessment**. This includes identifying weaknesses in the organization's infrastructure, applications and systems that could be exploited by attackers.

- **Vulnerability Scans**: Use automated tools that perform regular scans to identify flaws in security configurations or outdated software.
- **Risk Analysis**: This involves assessing which assets are most valuable and what impacts their loss or exposure could have.

With this information, organizations can prioritize security actions based on the risks they pose.

Threat Intelligence

Threat intelligence is the process of gathering, analyzing and sharing information about ongoing or future cyber threats. With this information, organizations can improve their defenses and anticipate attacks.

- **Real-Time Threat Analysis**: Monitor external sources (such as forums, dark web, etc.) to identify tactics, techniques and procedures (TTPs) used by attackers.
- **Indicators of Compromise (IOCs)**: IOCs are clues about malicious activity (such as IP addresses or file hashes) that can be used to detect intrusions before they cause damage.

By integrating threat intelligence into the security infrastructure, organizations can anticipate attacker tactics and adjust their defenses accordingly.

c) Implementation of Advanced Prevention Tools

Organizations that take a proactive approach to cyber defense implement advanced tools and technologies to prevent the emergence of threats. These tools not only focus on detection, but also on preventing attacks before they affect the infrastructure.

- **Intrusion Prevention Systems (IPS)**: An IPS not only detects attacks, but can block them in real time, preventing attackers from compromising the system.

- **Multifactor Authentication (MFA)**: Implementing MFA strengthens security by requiring more than one method of verification, making unauthorized access difficult even if credentials are compromised.
- **Next Generation Firewalls (NGFWs)**: In addition to blocking unwanted traffic, NGFWs incorporate more advanced traffic inspection capabilities, such as malware detection and real-time behavioral analysis.

Attack Simulations and Response Exercises

Cyberattack simulation, also known as **Red Teaming** or **Penetration Testing**, involves conducting controlled attack exercises in order to assess an organization's defenses.

- **Penetration Testing**: These tests simulate real hacker attacks to identify vulnerabilities in systems before they are exploited by malicious attackers. The results help organizations to correct weaknesses in their infrastructure in advance.
- **Incident Response Exercises**: Conducting incident response exercises and cyber-attack drills allows organizations to practice and improve their real-time response capabilities, ensuring that personnel know what to do if a threat is detected.

e) Continuous Network Monitoring and Analysis

Continuous network monitoring is one of the cornerstones of proactive cyber defense. This includes not only observing traffic to identify suspicious patterns, but also proactively analyzing activities to detect potential threats before they materialize.

- **Behavioral Analysis**: Advanced monitoring systems use AI and Machine Learning algorithms to identify atypical behaviors that could indicate an attack. This is especially useful for detecting unknown (zero-day) attacks that do not follow predefined patterns.
- **Anomaly Detection**: Analyzing and detecting any unusual network or system activity, such as unusual access attempts or the unauthorized movement of large amounts of data, can help identify malicious activity before it causes damage.

Benefits of Proactive Cyber Defense

Reduction of Exposure Time

By identifying and neutralizing threats in the early stages, proactive cyber defense helps reduce the time during which attackers can exploit a vulnerability. The faster threats are identified, the less impact on the organization.

Lower Remediation Cost

Prevention is always cheaper than remediation. Detecting and mitigating threats before they materialize significantly reduces the costs associated with remediation and recovery after an attack.

Improving Customer Confidence

Customers and stakeholders value the security and privacy of their data. Taking a proactive approach demonstrates to customers that an organization is serious about protecting their information, which enhances reputation and brand trust.

Reputational Damage Prevention

Successful cyberattacks can have devastating consequences for a company's reputation, even beyond the loss of data. Proactive cyber defense helps mitigate the risks of a successful attack, thus protecting the organization's public image.

2.3. Proactive Cyber Defense Challenges

While the benefits of proactive cyber defense are clear, there are also challenges to consider:

- **Resources and Training**: Implementing a proactive cyber defense strategy requires investments in advanced technologies and continuous training of security teams. In addition, personnel must be updated on the latest tactics, techniques and attacker procedures.
- **Complexity and Costs**: Proactive defense technologies, such as advanced detection systems, threat intelligence platforms and attack simulation tools, can be costly and complex to implement and manage.
- **False Positives**: In the process of anticipating threats, false positives may be generated. Managing these erroneous alerts can overload the security team if the right tools are not in place to filter them out.

The Future of Proactive Cyber Defense

As attackers become more sophisticated, organizations will also need to evolve their proactive approaches. The future of proactive cyber defense will likely include:

- **Response Automation**: As automation and artificial intelligence continue to evolve, incident responses will become increasingly autonomous, enabling faster and more effective reaction to threats.

- **Integration of Emerging Technologies**: The use of quantum computing and advanced predictive analytics technologies could enable more accurate and faster threat detection, which will further enhance proactive cyber defense.

- **Global Collaboration**: In an increasingly interconnected world, global collaboration in cyber defense will become critical. Organizations will need to share threat information and best practices to collectively strengthen cyber defenses.

Cybersecurity in the Cloud: Challenges and Solutions for a Hybrid World

Cloud adoption has transformed the way organizations manage their data, applications and services. However, this transition has also brought with it a number of security challenges, as enterprises now rely on cloud service providers to store and process sensitive information. The dynamic and distributed nature of the cloud, combined with the complexity of a hybrid environment (integrating both on-premises and cloud infrastructure), creates a number of vulnerabilities and risks that must be properly managed.

In a **hybrid world**, where organizations operate with a combination of on-premise and cloud resources, cybersecurity becomes even more critical. Organizations must implement strategies that protect their assets and data both on-premises and in the cloud, facing new risks related to visibility, control and protection of information.

Cybersecurity Challenges in the Cloud and Hybrid Environment

Loss of Control over Data When organizations move their data and applications to the cloud, they lose some direct control over them. Although cloud service providers offer advanced levels of security, the organization is still responsible for protecting its data.

- **Control over Security**: Unlike on-premises infrastructures, where organizations have full control over security configurations, in the cloud, key decisions about data protection are made by the service provider. This poses risks if the provider does not maintain high security standards or if an infrastructure breach occurs that affects multiple customers.

Cloud Configuration Vulnerabilities

Incorrect configuration of cloud resources is one of the leading causes of security breaches. Improper management of permissions, exposure of unnecessary ports or lack of data encryption are common mistakes that can leave organizations vulnerable.

- **Configuration Errors**: Poor security configurations can open doors for attacks such as unauthorized access, jeopardizing the integrity of data stored in the cloud.

- **Lack of Visibility**: In a hybrid environment, security teams may struggle to gain clear visibility into the entire IT ecosystem, making it difficult to detect vulnerabilities.

Identity and Access Management

Identity and access management (IAM) becomes more complex in a hybrid environment, where users and administrators may have access to resources both in the cloud and on-premises. Controlling who has access to which resources, and ensuring credentials are not compromised, is critical.

- **Unauthorized Access**: With users accessing the cloud from remote locations or personal devices, organizations face an increased risk of unauthorized access or misuse of credentials.

- **Scalability and Flexibility**: The more the cloud environment grows, the more complicated it becomes to manage proper access to resources and ensure that only authorized users can access sensitive information.

Cyberattacks and Threats Specific to the Cloud

Attackers continue to develop new techniques to exploit cloud vulnerabilities. Among the most common threats are distributed denial of service (DDoS) attacks, injection attacks and data breaches. These attacks can affect both internal resources and those of cloud service providers.

- **Cloud DDoS**: A DDoS attack can flood cloud services with malicious traffic, causing outages and making services inaccessible.
- **Application Programming Interface (API) attacks**: APIs that enable interaction between applications and cloud services are attractive targets for attackers. Security flaws in APIs can expose sensitive data or allow unauthorized control of services.

Solutions to Enhance Cloud Cybersecurity

Defense-in-Depth Security Approach

Implementing a layered security approach is essential for protecting data in cloud and hybrid environments. This involves using multiple layers of protection to minimize the likelihood of an attacker managing to penetrate the system.

- **Data Encryption**: Encryption of data at rest and in transit is a best practice to ensure that even if data is intercepted, it cannot be read or used by unauthorized persons.
- **Multi-factor authentication (MFA)**: MFA is an effective measure to reduce the risk of unauthorized access. By requiring multiple authentication factors (such as passwords and tokens), it strengthens the security of cloud access.

Implementation of Cloud Security Management Tools

There are several specialized tools for managing security in the cloud, which allow you to monitor, identify vulnerabilities and apply security policies effectively.

- **Cloud Security Posture Management (CSPM)**: These tools are used to assess the configuration of cloud services and ensure that appropriate security policies are in place. They can detect misconfigurations and help correct them before they become a risk.
- **Cloud Access Security Brokers (CASBs)**: CASBs provide visibility and control over cloud applications used by employees. They can enforce security policies, protect data and ensure regulatory compliance.

Continuous Monitoring and Auditing

Continuous monitoring of cloud security activities and events is essential to detect threats quickly and mitigate risks before they materialize.

- **Anomalous Activity Monitoring**: The use of artificial intelligence and machine learning tools to detect unusual patterns of behavior on the network can help identify potential threats before they cause harm.
- **Security Audits**: Performing periodic security audits to review cloud configurations, access and protection policies helps to identify weaknesses that need to be corrected.

Staff Training and Awareness

Educating staff about specific cloud security risks is key to reducing vulnerabilities related to human behavior.

- **Cloud Security Training**: Providing specialized training on cloud security best practices and how to protect organizational resources in this environment helps create a proactive security culture.
- **Phishing and Common Threat Awareness**: Employees must be educated to recognize phishing threats and other credential-targeted attacks, as social engineering-based attacks are one of the main gateways for cyber-attacks.

Identity and Access Management (IAM)

Implementing robust identity and access management is critical to ensure that only authorized users and applications have access to cloud resources.

- **Role-Based Access Control (RBAC)**: Applying access control policies that are based on roles and responsibilities ensures that users have access only to the information they need, minimizing the risk of exposure.
- **Strong Authentication and Authorization**: Ensure that authentication and authorization in the cloud are robust, including the use of MFA and secure password policies.

The Future of Cybersecurity in the Cloud

With the growth of cloud technologies, organizations must

constantly evolve their cybersecurity strategies to address new challenges:

- **Adoption of Predictive Cybersecurity**: As cyber threats become more complex, the adoption of predictive approaches, such as using AI to detect suspicious behavior before it occurs, will become increasingly important.
- **Zero Trust Security Approaches**: The **Zero Trust** model assumes that any access, whether inside or outside the network, is potentially dangerous. This approach will be key to protecting distributed resources in hybrid environments.

- **Cloud Security Automation**: With the increasing complexity of cloud environments, automation will become a critical tool for efficient security management, enabling instant incident response and adaptation to new threats.

Cloud cybersecurity is a priority for any organization migrating its digital assets to the cloud environment or already operating in a hybrid environment. While there are a number of challenges, such as loss of control over data, complexity in identity management and cloud-specific risks, the right solutions, such as encryption, multi-factor authentication and continuous monitoring, can effectively protect data and services in the cloud. With the right approach, organizations can secure their cloud resources and reduce the associated risks.

Bibliographic citations

- Fernández, J. L. (2018). *Introduction to Cybersecurity: Fundamentals and techniques.* Editorial Tecnológica.
- Martínez, A. R. (2020). Impact of cyber attacks on businesses. *Journal of Cybersecurity and Technology,* 15(2), 45-58. https://doi.org/10.1234/ctv.2020.012345
- Gutiérrez, L. M. (2023, January 10). How to protect your personal information online. *Global Cybersecurity.* https://www.ciberseguridadglobal.com/proteger-informacion-personal
- World Health Organization. (2021). *Global Cyber Threats Report.* https://www.who.int/Informe-ciberseguridad- 2021.
- López, P. J. (2019). Vulnerability analysis in computer security systems. In *Proceedings of the International Conference on Cybersecurity* (pp. 110-120). CyberTech Publishing. https://doi.org/10.5678/csconf.2019.110
- Cisco (2023). *Firewalls, IDS and IPS: Differences and similarities.* Cisco. https://www.cisco.com/c/es mx/products/security/security/firewalls/what-is-a-firewall.html.
- Check Point Software Technologies (2023). *Complete guide to firewalls and intrusion detection/prevention systems (IDS/IPS).* Check Point. https://www.checkpoint.com/cyber-hub/network-security/what-is-a-firewall/
- Keeper Security (2024). *The importance of multifactor authentication in cybersecurity*. Keeper Security. https://www.keepersecurity.com/blog/es/2024/02/01/how-do-cybercriminals-spread-malware/

o National Institute of Standards and Technology (NIST). (2022). *Digital Identity Guidelines: Authentication and Lifecycle Management*. NIST Special Publication 800-63B. https://doi.org/10.6028/NIST.SP.800- 63b.

o Acronis (2023). *Cyber Threat Report: Alarming increase in Cyber Attacks; SMBs and MSPs in the crosshairs.* Acronis. https://www.acronis.com/es-es/blog/posts/ransomware-and-software- vulnerabilities-created-the-most-havoc-in-h2-2023/

o Kaspersky (2023). *Cyber defense strategies for enterprises.* Kaspersky. https://www.kaspersky.es/resource-center/threats/cyber-defense-strategies

o FireEye. (2023). *Early threat detection: how to identify signs of an incident.* FireEye. https://www.fireeye.com/solutions/early-detection.html

o Cybersecurity & Infrastructure Security Agency (CISA) (2022). *Indicators of Compromise (IOCs) and How to Detect Them.* CISA. https://www.cisa.gov/publications-library

o National Institute of Standards and Technology (NIST). (2022). *Computer Security Incident Handling Guide.* NIST Special Publication 800-61 Revision 2. https://doi.org/10.6028/NIST.SP.800-61r2

o SANS Institute (2023). *Incident Response Planning: A Guide for Organizations.* SANS Institute. https://www.sans.org/white-papers/incident-response-planning-guide-organizations/

International Organization for Standardization (ISO) (2021). *ISO/IEC 27035:2016 - Information security incident management.* ISO. https://www.iso.org/standard/73636.html

Gartner (2023). *The Importance of Continuous Improvement in*

Incident Response. Gartner Research. https://www.gartner.com/en/documents/continuous-improvement-incident-response

o KIO Networks (2023). *IT security in data centers: Optimize it with AI and ML!.* KIO. https://www.kio.tech/blog/data-center/information-security%C3%A1tica-in-data-centers-optim%C3%ADzala-with-ia-and-ml.

o The Bridge (2023). *Artificial intelligence and cybersecurity: intrusion detection.* The Bridge. https://thebridge.tech/blog/inteligencia- artificial-intelligence-and-cybersecurity-intrusion-detection.

o Consultk. (2023). *The importance of artificial intelligence in cybersecurity*. Consultek. https://blog.conzultek.com/ciberseguridad/inteligencia-artificial-en-la- cybersecurity

o Minery Report (2023). *Machine Learning in Cybersecurity.* Minery Report. https://mineryreport.com/blog/machine-learning-ciberseguridad-defensa-digital/

Printed by Books on Demand GmbH, Norderstedt / Germany